Grandpa's John Deere Tractors

Roy Harrington

Published by the
American Society of Agricultural Engineers
2950 Niles Road, St. Joseph, Michigan

Photo Credits

John David King/Photo Designs: front cover.
Deere & Company: 1, 4, 5, 6a, 6b, 8a, 8b, 9, 10a, 10b, 12a, 12b, 13, 14a, 14b, 15, 16b, 18a, 18b, 19, 20a, 20b, 21, 22a, 22b, 23, 24a, 24b, 25, 26, 27, 28, 29, 30a, 30b, 32a, 32b, 33, 34a, 34b, 35, 36a, 36b, 38a, 38b, 39, 40a, 40b, 41a, 41b, 42b, 42c, 42d, 43a, 43b, 43c, 43d, 44a, 44b, 45, back cover.
Trademarks by Permission of Deere & Company: 3.
Whitey's Ice Cream: 7. Derwood Heine: 11.
Robert E. Hay: 16a. Dwight Emstrom: 17. NASA: 27.
Reprint by Permission of Texas Instruments: 29.
Courtesy of International Business Machines Corporation, unauthorized use not permitted: 31, 37. THE ERTL CO., INC.: 42a.

Front Cover

Mike Williams of Clinton, Iowa, and his 1936 John Deere Model "D" Tractor.
Mike was president of Deer Valley Collectors in 1995-1996.

Grandpa's John Deere Tractors

Editor: Richard Balzer
Project Manager: Melissa Carpenter Miller

Library of Congress Cataloging-in Publication Data

Harrington, Roy
Grandpa's John Deere Tractors

Summary: A farm boy learns from Grandpa how John Deere tractors and implements made farm life easier. His parents update him on more recent John Deere products.
1. Farm mechanization — History — Juvenile literature. 2. John Deere tractors — Juvenile literature. 3. Agricultural machinery — Juvenile literature. 4. Farms — Juvenile literature.
[1. Farm mechanization — History. 2. John Deere Tractors. 3. Farms.] I. Title.
1996 631.3

ISBN 0-929355-82-2
LCCN 96-86572

Printed in the United States.

Table of Contents

1876

1936

1950

1956

1968

John Deere demonstrates the self-scouring plow he developed to interested farmers. An old saw blade was used to make his first steel plow in 1837. His blacksmith shop was in Grand Detour, Illinois.

1837 – John Deere Helps Change Farming

Grandpa knows everything about John Deere Tractors!

Sometimes, Mom lets me get off the school bus early at Grandpa Turner's farm. I run to the house, drop my school bag in the kitchen, hug Grandma, grab a bag of fresh-baked chocolate-chip cookies, and head for Grandpa's shop. Soon Grandpa and I are talking John Deere tractors.

"Grandpa, did you ever talk to your dad and granddad about John Deere equipment like we do?"

"I certainly did, Jim. My Grandpa Robinson told me the 1800s were an exciting time for change on the farm. John Deere's steel plow had made it much easier for farmers to plow the prairies for the first time. McCormick invented a reaper in 1831 that cut and gathered grain. Forty years later, Deering developed a binder that also tied the grain in bundles. A few years later, in the

In 1890, this giant steam engine was strong enough to pull a combination of ten John Deere plows. Two men rode on the machine to raise each plow at the end of the field. John Deere had moved his plow factory to Moline, Illinois, in 1848. The Mississippi River drove water wheels to power his factory.

1880s, J.I. Case became the major manufacturer of threshers. These machines removed the wheat and oats from the straw, cleaned the grain, and stacked the straw. J.I. Case threshers were originally powered by horses. Later J.I. Case made steam engines to power his threshers. His steam engines became the most popular ones sold.

"Grandpa Robinson said that in the early 1800s most farm work except plowing and hauling was done by hand. Farming took so much work before these inventions that more than half the people in the United States had to live on farms just to grow enough food to feed the rest. So, farmers welcomed each new machine that made farm work faster, easier, and better. By 1900, farmers were harnessing horses for most heavy farm work. But some larger farmers bought steam engines to run their threshers and to help thresh their neighbors grain.

"Important changes were also happening off the farm. Discovery of gold in California in 1848 started the gold rush. Twenty years later, the Union Pacific Railroad extended its tracks across the continent, from ocean to ocean. For sure, Jim, the 1800s produced many famous inventors and great inventions."

From the long-ago inventions Grandpa told me about, I think the most important were photography, sewing machines, typewriters, electric motors, electric lights, telephones, radios, bicycles, and cars.

Froehlich tractor, 1892, 1-cylinder Van Duzen gasoline engine, 1-speed ***transmission****.* (Look in the glossary for boldface words.) The Froehlich was the first gasoline tractor to have a gear for reverse. The driver had excellent visibility as he sat or stood at the front.

1892 – First Tractor Sold to Run Thresher

The Froehlich tractor successfully operated a large J.I. Case threshing machine for a 50-day run in 1892. It eliminated the fire used in steam engines. This reduced startup time in the morning and removed the serious danger of setting fire to the straw stack.

"Grandpa, when were tractors invented?"

"It's a little hard to say, Jim. Many people were experimenting, looking for something better than a steam engine. The steam engine had lots of power and pulled heavy loads, but it required more than an hour each morning to build the fire and get up steam. In 1878 Otto demonstrated an internal combustion engine, the type we have in cars today. These gasoline engines started replacing stationary steam engines.

A girl enjoys a Mississippi Mud ice cream cone in John Deere's hometown, Moline, Illinois. Ice cream cones were first sold in 1904 at the St. Louis World's Fair.

"In 1892, about the time my father was born, John Froehlich built and sold the first two successful gasoline tractors in the United States. These tractors were designed to power threshing machines and to pull them from one farmer's field to another. But they weren't designed to pull plows. Although Froehlich's tractors had a successful threshing season, the buyers were unhappy and returned both of them.

"Froehlich was still convinced gasoline tractors were going to replace steam engines. He and some others started the Waterloo Gasoline Traction Engine Company in late 1892. Many years later, that company was sold to become the John Deere Tractor Company. That's why these early tractors didn't have the John Deere name.

"You see, tractors have families and family trees just as people do. For instance, my last name is different from yours because I'm dad to your mother rather than your father — but we're still on the same family tree. Over the years, the John Deere family of tractors kept growing with different sizes like bigger and smaller brothers and sisters. You might even think of the yellow John Deere industrial tractors as cousins of the green farm tractors."

"Were steam engines making farming a lot easier when your dad was born?"

"Not really. Steam engines helped some farmers with threshing and plowing but most farm work was done by horses and the farm family. My dad only went to grade school six months each year because the children were needed for farm work during the growing season. But by the time I went to a one-room country school our help wasn't needed as much on the farm, so we went eight months. Nowadays you have to go nine months just to keep up with advances in technology."

"Farming's getting easier, Gramps; school's sure not!"

Model "N" Waterloo Boy Tractor, ***1917-1924, 25 horsepower****, 2 cylinders, 2-speed transmission.* The engine started on gasoline and then switched to cheaper kerosene to run. The two forward speeds of 2 1/4 and 3 miles per hour were similar to those of horses and mules.

The Waterloo Boy Tractor was bought for plowing by more than 8,000 farmers before the company was purchased by **Deere & Company** in 1918. The cleats on the rear wheels and the rugged transmission made it suitable for plowing and other **drawbar** work.

1918 – Deere & Company Buys Tractor Factory

"Grandpa, when did John Deere start going from horse-drawn to tractor-powered equipment?"

"Some of the John Deere **sales branches** started selling other brands of tractors in 1910. Sales of steam engines by J.I. Case hit their peak in 1912 and soon declined rapidly in favor of tractors. Many people recognized the market potential, so more than 100 companies made tractors from 1916 to 1922. A few individuals in John Deere designed and built tractors but none appeared to be the right answer. Therefore, the company decided it was better to enter the market with a farm-proven design. So Deere & Company bought the Waterloo Gasoline Engine Company in 1918. More than 26 million horses and mules were on farms when tractors began to replace horses. John Deere had gotten into the market just in time to grow with it."

Henry Ford introduced the compact 18-hp Fordson tractor with cast-iron frame in 1918. Prices as low as $375 led to annual sales of more than 100,000 tractors in some of the next 10 years.

"Was the Waterloo Boy a good tractor?"

"It was for its day. Many tractors built then were not dependable. Because of this, the state of Nebraska set up a test for tractors to protect the farmers in their state. The Waterloo Boy was the first tractor to pass their endurance tests. It was also advertised as 'The Original Kerosene Tractor.' The tractor was started on gasoline. When it had warmed up, it was switched to burning lower-cost kerosene. Farmers liked this saving in fuel costs."

"I notice most Waterloo Boy Tractors are green, red, and yellow. Why is John Deere equipment green?"

"John Deere uses green because it is the color of crops growing on farms like corn, wheat, and grass."

Grandpa didn't tell me, but I think yellow wheels are used because they look like flowers, maybe dandelions or daffodils.

"Grandpa, what was the first tractor your Grandfather Robinson owned?"

"He never owned one, Jim. By 1920, he had a telephone, a radio, and an early Model T Ford car but he never said he wanted a tractor. When he retired he still thought horses were more dependable than tractors. Most farmers either agreed with him or thought tractors were too high priced because less than 15% owned tractors in 1930."

Model "D" Tractor, 1924-1953, 27 hp, 2 cylinders, 2-speed transmission. The Model "D" added a third important way of delivering power to implements. Its power take-off (PTO) shaft powered grain binders and some other equipment that had used ground-drive wheels with horses.

1924 – Waterloo Makes a John Deere Tractor

"Grandpa, why did you choose the Model "D" (shown on the front cover) as the first tractor you restored?"

"To me, the Model "D" represents great engineering. It was the first tractor designed by John Deere engineers instead of blacksmiths. The design was so good the Model "D" was made for 30 years — longer than any other John Deere tractor. It also set the basic design for most later 2-cylinder tractors.

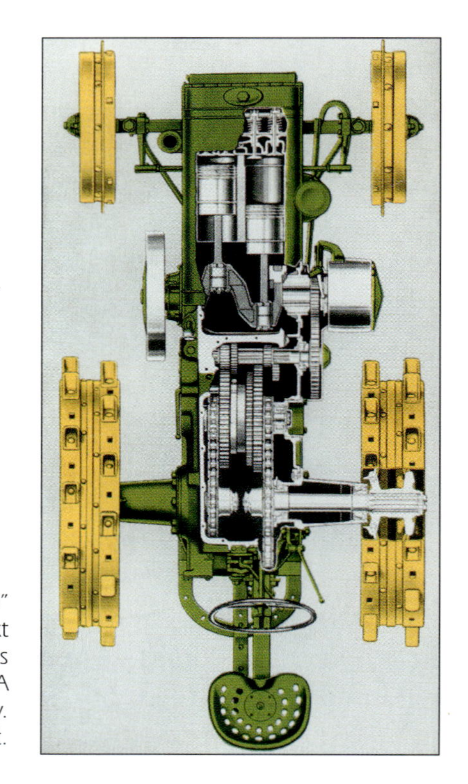

The simple, clean, compact design of the Model "D" was to be followed by John Deere for the next 35 years. The horizontal, crosswise engine had its flywheel on the left and the belt pulley on the right. A hand lever operated the clutch inside the belt pulley. Speeds were changed by sliding gears on a cross shaft.

International Harvester made the 18-hp Regular Farmall in 1924-1932. Its tricycle design permitted easy cultivation of crops planted in rows like corn and cotton. The tricycle design was used for most farm tractors for more than 30 years.

"John Deere engineers from the Waterloo factory visited farmers who owned Waterloo Boy tractors and asked them what needed to be improved. They wanted more power in a smaller tractor that would be easier to turn at the end of the field. Enclosed engine and drive parts were requested so less time needed to be spent with the oil can servicing the tractor. Farmers asked for a better place for the driver, away from the heat of the engine. Customers asked for a way to power implements in the field just the way the belt pulley powered stationary equipment at the farmstead. A power take-off (PTO) shaft between the tractor and the implement met this requirement.

"I think John Deere's original success with plows resulted from him knowing what farmers needed and then giving them a quality product. John Deere engineers have always gone out and talked to farmers about what they like about their current equipment and what they want in the future. They get a lot of their good ideas for better equipment from farmers, Jim.

"Farmers liked the slow putt-putt, miss-miss sound of the Model "D" engine. They found the engine was good at 'lugging' to pull the tractor through tough spots without stalling. That putt-putt sound of the John Deere 2-cylinder engine has been music to the ears of many farmers for more than 70 years. Even though these tractors haven't been made for more than 35 years, many are still working on farms.

"But, down deep, there was a personal reason I chose the Model "D" as the first I'd restore: it was made about the time I was born. Some people think we're both old, but we're still going strong."

1928 – A General Purpose Tractor is Added

Model "GP" Tractor, 1928-1935, 20 hp, 2 cylinders, 3-speed transmission. The "GP" had an arched front axle to clear corn when used with the 3-row cultivator shown. The "GP" was the first tractor to offer a fourth way to use power, a mechanical power lift to raise mounted cultivators, planters, and mowers.

"Grandpa, why did John Deere make a "GP" Tractor if the Model "D" was so good?"

"The Model "D" is sometimes called a standard or a wheatland tractor. Many first buyers of tractors grew wheat on large farms in the west. The "D" met their needs but did not have enough clearance under its axles to cultivate corn. The General Purpose tractor was advertised as 'The two-plow tractor of standard design that plants and cultivates three rows at a time.' Cornbelt farms were smaller so the "GP" was smaller, lighter, and less expensive than the "D"

Model "GP" Tractor with a mounted mower powering an overshot hay stacker. The tractor umbrella is folded so the driver can look back to see when to stop. At that time, it was more common for a horse to pull the rope to lift the hay from the ground and drop it on the stack for the man to spread.

Model "GP" Wide-Tread Tractor, 1929-1933, 20 hp, 2 cylinders, 3-speed transmission. This tricycle tractor was designed for farmers that wanted 2- or 4-row equipment instead of the 3 rows matched to the earlier "GP" design. Both tractors had individual rear-wheel brakes for short turns at the end of the field.

Tractor. One of the ways they made it cheaper was to use a simpler flat-head engine. All other Waterloo tractors have used valve-in-head engines because they are more fuel efficient."

"Grandpa, how did you make hay when you were growing up?"

"It wasn't easy, Jim. The hottest weather of the summer seemed to always arrive when we cut hay. I grew up on a cattle farm so we had to grow a lot of hay for them and some for the horses and sheep. When our three barn lofts were full of hay, we started stacking the rest of it outside. My first job was to lead the horse, Dinah, on the rope that pulled the hay up into the barn. Dinah was also used to pull the hay up for stacks. The next job I got when I was older was using the sulky rake to gather hay into windrows. The next year I was old enough to drive one of the horse-drawn mowers. By then I had enough muscle I could also be one of the two pitching hay up on the wagon. My final promotion was to spread the hay in the barn loft as it came in. This was the hottest, dirtiest job in haying.

"In those days, farm folks and others who had to pump all their water by hand only took bathes on Saturday nights. We had it better than most farmers because we had running water in our house. To get this, we pumped water up into a horse tank in the girls' bedroom. We had enough water for the toilet and to wash our hands and face any time. After we got hot and dirty from haying we would go jump in the pond to cool off and get rid of the dust. The cows, who thought they owned the pond, would get out when they saw us coming. They didn't want to associate with anyone so dirty. The pond was far enough from the road we just took our clothes off on the bank and swam without suits."

Unstyled Model "A" Tractor, 1934-1938, 23 hp, 2 cylinders, 4-speed transmission. The "A" kept the basic John Deere 2-cylinder design but added an optional lower-cost fuel called distillate. The engine continued to be started by turning the flywheel by hand. Petcocks released piston compression for easier starting.

1934 – More Power and Versatility

When it's warm enough, Grandpa and I fish a lot. It's another time when we talk tractors because we don't worry about scaring the fish away like some folks do. "Grandpa, did your dad ever own an unstyled Model "A" like the one my dad restored?"

"No, Jim, my dad said we didn't have enough money to buy a tractor because of the depression in the '30s. Some of our neighbors were buying tractors, but Dad bought another farm so we could keep more cattle. He was also spending money on two things that farmers had not purchased before. He started buying **hybrid seed corn** and commercial fertilizer. We had the first demonstration plot for hybrid corn in our county. The university was promoting hybrid corn, but farmers had to see the yield difference on the farm of someone they trusted before they would spend extra money for hybrid seed. Yields of the new hybrids really jumped when fertilizer was used.

"Tractor design was changing as fast as other changes on the farm. Allis-Chalmers was the first to sell a tractor on rubber tires. It took some time to convince farmers that tractors with rubber tires would do as much work. Even then, many farmers chose steel wheels because they cost less and lasted longer.

Unstyled Model "A" Tractor shown mowing a roadside with a No. 5 Mower. The "A" added more attaching points for mounted implements like mowers, cultivators, plows, and planters. These tools could be lifted by a tap of the heel to operate the new hydraulic lift.

"The unstyled Model "A" had two firsts later adopted by other tractor manufacturers. A hydraulic lift on the tractor took the muscle work out of lifting cultivators and other mounted equipment. And the width between the centers of the rear wheels could be adjusted from 56 to 84 inches. This was done by sliding the wheels along the splined rear axles. Corn rows were about 40 inches apart to match the width of horses, but farmers liked to grow potatoes and other vegetables in rows narrower than this.

"Engineers must have spent many hours driving the experimental tractors because they certainly improved the driver's area. The operator was moved up out of the dust and forward for a better ride. He had a good place for his feet when sitting or standing. The hood of the tractor was narrow for excellent vision when cultivating. Some farmers had used a tractor for their heavy work but kept horses for jobs like planting, cultivating, and mowing. With as much drawbar power as the "D" or "GP," here was a tractor that could replace all the work horses on the farm."

Dad and his restored unstyled Model "A" in our always-neat shop with Karen and me. I like the way Grandpa restores tractors. He has tractor parts scattered all over the shop floor just like our toys in the family room. We have to pick up our toys every night, but Grandpa only picks up once a week to sweep.

Styled Model "B" Tractor, 1939-1952, 16 hp, 2 cylinders, 4-speed transmission. A couple of years after the styled "B" was introduced, it was equipped with a 6-speed transmission to provide higher road speeds with rubber tires. Electric starting was originally an option but became standard on later tractors.

1939 – Styling Announces Improvements

"Grandpa, what were the biggest changes in farming when you were in high school?"

"I was in high school in the early '40s when lots of changes came to many farms, Jim. The **REA** brought electricity to most farms. We finally got our dirt road graveled. But the biggest change to me was the day in 1941 when the John Deere dealer delivered our new Model "B" Tractor. It had everything — electric starter and lights, rubber tires, and a road gear of 10 miles per hour.

The Model "B" was a popular power source for the 5-foot 12A Combine introduced in 1940. The 12A became the most popular John Deere combine ever made with more than 116,000 built in 13 years.

More than 300,000 Ford-Ferguson 20-hp tractors were made between 1939-1947. Their new 3-point hitch established the best way to hitch and control rear-mounted implements. The basic principle of that hitch design continues to be used on most tractors today.

I could pass any team of horses on the road. It looked as great as any car at that time. When you look back at the unstyled tractors, like your dad's restored Model "A", they look like they are still in their underwear."

"Aw, c'mon, Grandpa, Karen and I don't look crude in our underwear!"

"You got me there, Jim. Anyway, Dad got a plow, a cultivator, a mower, and a field cultivator with the new tractor. We just kept using our horse-drawn disk, spike-tooth harrow, and manure spreader behind the tractor. And Dad kept a team of horses for planting and a few other jobs.

"A John Deere ad said, 'While the Model "B" Tractor weighs less than two good horses, it will do the work of four.' Farmers found this to be true, making the Model "B" the most popular tractor John Deere ever built. More than 306,000 of them were made in 1935 through 1952. Sales of the Model "A" were only slightly less than that figure.

"An equally big change came the following year when Dad got a 12A Combine to harvest and thresh our grain. I liked the independence it gave us. But I missed all of our neighbors working together in a threshing ring so that everyones grain was threshed. The man who owned the threshing machine would go from farm to farm to thresh its wheat and oats. Some of the farmers brought their horse-drawn wagons and others came to pitch the bundles of grain up onto the wagons. Threshing was another hot, dirty job like making hay. When I was your age, I worked in the grain wagons to see they got full and nothing spilled on the ground. When I was a little older I rode a horse through the fields, delivering fresh water to all the workers. We used a gallon glass jug with a corn cob for a cork. The jug was wrapped with a wet sack to keep the water cool. The best part of threshing was the huge meal served by the farm wives. The meals always ended with a choice of scrumptious pies."

1939 – Smaller Tractors for First-Time Owners

Model "H" Tractor, 1939-1947, 12 hp, 2 cylinders, 3-speed transmission. It had the family appearance of its big brothers at Waterloo. It differed by having the belt pulley and clutch on the camshaft instead of the crankshaft. For road travel, a foot throttle increased the engine speed from 1,400 rpm to 1,800 rpm.

"Grandpa, did John Deere ever make farm tractors smaller than the Model "B" you first drove?"

"Yes, Jim, three different factories made tractors smaller than the "B" Tractor. That's because farms come in many sizes. Many of the farmers using only two to four horses or mules did not think they could afford to buy the Model "B" and the equipment to use it. Although some machines designed for horses could be hitched to a tractor and controlled by the driver, other machines like cultivators and mowers took a second person. The smallest model ever made at Waterloo was the "H" Tractor. It was a simplified version of the three larger general purpose or row-crop tractors. It was a perfect match for a two-row cultivator.

Model "L" Tractor, 1937-1946, 9 hp, 2 cylinders, 3-speed transmission. The "L" differed from Waterloo tractors by having smaller rear wheels, a foot clutch, and a vertical, in-line engine.

Model "M" Tractor, 1947-1952, 18 hp, 2 cylinders, 4-speed transmission. The "M" also had a foot clutch and a vertical, in-line engine. While still a one-row cultivating tractor, it had twice the power of the "L" Tractor.

"Before Waterloo introduced the "H," the John Deere Wagon Works in Moline entered the market with an even smaller model, the "L" Tractor. It weighed in at a mere 1,515 pounds. The engine was slightly offset to the left and the operator to the right for better vision to cultivate one row. It powered the same size implements as a team could, such as a 12-inch plow, a one-row cultivator, and a 5-foot mower. However, at the end of the day the farmer had done more work because the tractor didn't have to be rested like horses.

"In 1947, the "L" and "H" were replaced by the Model "M," made in a new factory at Dubuque, Iowa. It was a more conventional design than the "L" and had a new John Deere engine. It retained the offsets of the Model "L" for vision to cultivate a single row. Also, like the "L," it used drop housings for the final drives to get crop clearance with smaller rear wheels. Implements on the Model "L" had to be manually lifted while the "M" offered hydraulic lift and control. A similar two-row Model "MT" was introduced in 1949.

"These three tractors helped in the final push to get farmers to switch to tractors from horses and mules. Tractors on U.S. farms grew from 1 million in 1932 to 2 million in 1943, 3 million in 1949, and 4 million in 1953. Horse and mule population declined from more than 26 million in 1920 to less that 8 million in 1950. Each additional tractor replaced an average of four horses."

Model "R" Diesel Tractor, 1949-1954, 43 hp, 2 cylinders, 5-speed transmission. The Model "R" used a small 2-cylinder gasoline engine to start and warm up the larger diesel engine. The "R" was the first John Deere tractor to have a separate clutch for the PTO to make it independent of tractor travel.

1949 – John Deere's First Diesel Tractor

"Grandpa, when did John Deere first make a big diesel tractor?"

"The Model "R" Diesel was introduced in 1948 for the 1949 production year as the first John Deere tractor with more than 40 horsepower. Weighing in at 7,400 pounds, this muscular tractor had almost five times the weight and power of the Model "L" Tractor. Although the drawbar power of the Model "D" had doubled between 1924 and 1940, farmers were looking for even more power and weight to pull wide implements. The wheat and rice farmers didn't need to cultivate crops in rows so still preferred the standard fixed-tread design. The "R" set new fuel economy records in Nebraska Tractor

This Model "R" pulls a tandem CC Field Cultivator, 23 feet wide. Hydraulic cylinders raised the cultivator sections at the end of the field. In 1945, John Deere was the first tractor manufacturer to offer remote hydraulic cylinders to lift and control drawn implements.

50 Tractor, 1952-1956, 26 hp, 2 cylinders, 6-speed transmission. This replacement for the Model "B" had more drawbar power than the Model "D" during its first 10 years. Its PTO and its hydraulic system were each independent of tractor travel. It is shown operating a 227 Corn Picker and pulling a wagon.

Tests. This small appetite with the lower cost diesel fuel meant a real savings for farmers who used their tractor many hours each year."

"Grandpa, what model replaced the Model "B" you grew up on?"

"The 50 was an even better tractor than our "B," Jim. It had a dual carburetor that improved fuel efficiency and gave it more power. John Deere also introduced a quicker and easier way to adjust rear wheel tread. A rack-and-pinion was built in to help move the wheels in or out. In 1954, John Deere was the first tractor maker to offer factory-installed power steering on row-crop tractors. Another John Deere exclusive, that I wish our "B" had, was Roll-O-Matic front wheels. They cut bumps in half by having one wheel go down by the amount the other went up over a bump. This was one of several inventions that Deere & Company bought from farmers to use on its equipment."

"Grandpa, what was the last use for horses on farms in the Midwest?"

"Picking corn by hand with a team of horses was one of the last field jobs to be mechanized. The team pulled the wagon along beside the man picking the corn. A well-trained team would respond to voice commands, just like the latest computers. They would move forward in the row with 'Giddyup' and stop with 'Whoa.' Harvesting corn by snapping each ear is much faster than harvesting wheat by hand cutting. It was also done from an erect position in cool weather. My dad could pick and unload 100 bushels of corn in a day."

1956 – Less Driver Work, More Tractor Power

620 Tractor, 1956-1958, 41 hp, 2 cylinders, 6-speed transmission. Shown with 14T Baler throwing a small bale into a wagon. This replacement for the 60 had 17% more power. A new 3-point hitch gave improved control of mounted implements. A bright splash of yellow paint on the hood announced the new models.

"Grandpa, when did making hay get easier than when you were growing up?"

"The first step was a switch from the loose hay we handled in barns and stacks to baled hay, Jim. There were stationary balers when I was your age but they didn't save labor. The big savings came when we got balers that picked up the hay in the field. John Deere introduced the first baler to automatically wrap and tie wire around bales, to hold them tight together, in 1946. This was

730 Diesel Tractor, 1958-1961, 50 hp, 2 cylinders, 6-speed transmission. Shown with tandem 494 Planters covering 8 rows. An interesting feature of the 2-cylinder 730 Diesel was its high-speed V-4 starting engine. Farmers thirsting for more power made the 730 the best seller of the six power sizes of 30 series tractors.

Here's a 10 Corn Head on a 45 Combine harvesting corn. Different from tractor-powered corn pickers, the design of the corn head cut shelling losses in half and reduced corn harvesting accidents.

followed in 1954 with the twine-tie 14T Baler. These balers still required a minimum crew of three to pick up the bales from the field and store them in the barn.

"The next big change came in 1957 when John Deere was the first to sell a bale thrower to automatically toss bales into the trailing wagon. The wagon was then taken to the barn where elevators took the bales up into the barn and distributed them in the loft. Finally, here was a one-man system to bale and store hay in one day before it got rained on."

"Grandpa, what machine besides tractors changed farming the most after you started farming?"

"The corn head for combining corn was the most important change, partly because corn is the most important crop in the United States. John Deere was again first on the market, in 1955, with a corn head for the 45 Combine. Corn pickers had taken the hard work out of picking ear corn. However, ear corn is difficult to get out of the wagon and into the crib for storage. It is even more difficult to get out of the crib to feed as ears or to shell for feeding or selling.

"Combining corn changed all this because shelled corn flows easily so it can be mechanically conveyed into and out of storage. There are many other advantages over corn pickers. Harvest can start earlier because the corn can be hauled directly to be sold or dried on the farm. The combine can operate in softer, muddier fields because the weight is concentrated over the drive wheels instead of the small front tires of a tricycle tractor with mounted corn picker. Combining corn doesn't tie up a tractor that may be needed for other jobs. Finally, 2-row mounted corn pickers are too slow. Most corn heads sold in 1995 harvested either 6 or 8 rows."

3010 Tractor, 1961-1963, 59 hp, 4 cylinders, 8-speed synchronized transmission. This transmission permitted shifting on-the-go without clashing gears. Tractor shown with a 406 Lister Planter. With more power than a 730 Tractor, the 3010 was able to operate 4-row planters at higher speeds.

1961 – A New Generation of Power

I watched Grandpa parade his Model "D" at the Mississippi Valley Fair. Deer Valley Collectors drove more than 30 restored John Deere 2-cylinder tractors around the fair. Each collector seemed to think 2-cylinder tractors were the best. After the fair I asked, "Grandpa, what did you think when John Deere stopped making 2-cylinder tractors and started making 4- and 6-cylinder tractors?"

"I didn't like it, Jim. My dad and I had always farmed with 2-cylinder John Deere tractors. We liked their lugging ability, fuel economy, simplicity, and

5010 Tractor, 1963-1965, 121 hp, 6 cylinders, 8-speed synchronized transmission. Shown with a 2-row 12 Forage Harvester and 115 Chuck Wagon. This was the first 2-wheel-drive tractor a farmer could buy with more than 100 horsepower on the drawbar, enough to cut two rows of heavy corn when the fields were soft.

The first observation farmers made when they climbed on the new tractors was how much more user-friendly they were. The adjustable seat was very comfortable and the controls were easy to reach and operate.

dependability. I liked being able to change the clutch myself without sending it to the dealer to take the tractor apart in the middle. Its putt-putt sound was, and still is, music to my ears.

"Fred, my neighbor who always had to have the latest and biggest of everything, got a 4010 Tractor in 1961. Of course, he invited me over to look at it and drive a round with his new mounted plow. I was amazed at the amount of power it had and how easily it handled. I started looking at it more carefully and asking him questions. The tractor was not only a big change from the 2-cylinder John Deeres, it was a distinct improvement over any other tractor on the market.

"I liked the easy, shift-on-the-go 8-speed synchronized transmission. With the added power of the tractor, it was practical to plow and work with other implements at higher speeds. Having 8 speeds, it was possible to match the tractor's power with the implement's most productive operating speed. The hydrostatic power steering and the hydraulic power brakes made the tractor easier to handle than tractors half that powerful. The new 3-point hitch worked better with large implements than any tractor I had used.

"I still questioned whether a tractor this new and revolutionary could be dependable like my 2-cylinder tractors. Fred said he knew from experience John Deere tested its equipment thoroughly before selling it to farmers. He also said when problems had arisen his dealer had always corrected them. He had heard that John Deere engineers did lots of field testing during the seven years they worked on these new tractors before making and selling them."

The 3010 and 4010 had introduced the most advanced hydraulic system on the market. Its new pump provided instant power when needed for steering, brakes, 3-point hitch, and remote cylinders. The succeeding 3020 and 4020 also had a power differential lock to control spinning in slippery conditions.

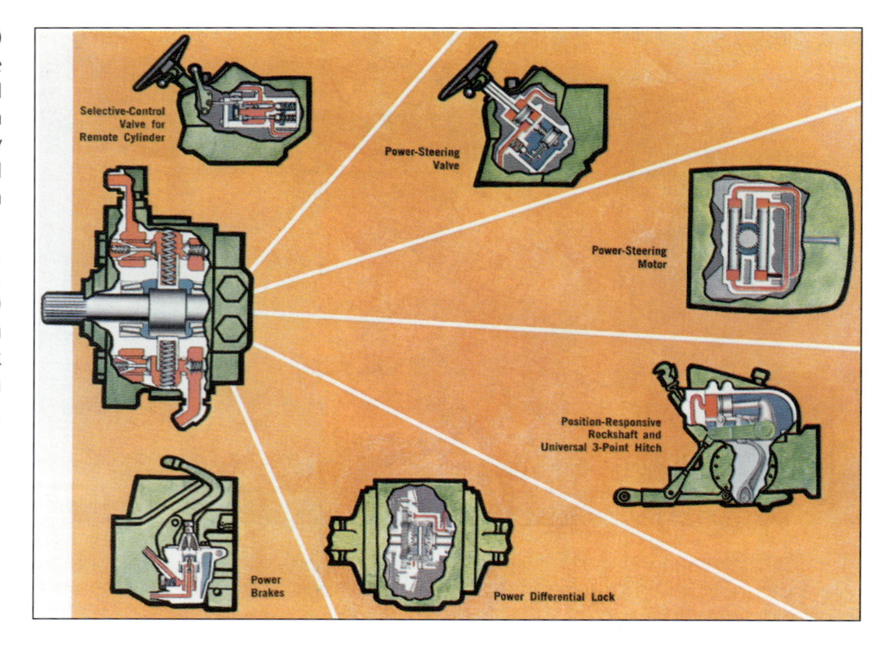

1964 – A Best-Seller, the 4020

"Grandpa, what was the first new tractor that you bought?"

"The John Deere 4020 Diesel Tractor (shown on the title page) is the first new tractor I owned. I started out with a used Model "B" and later bought a used 720 Diesel at a farm sale. I had a chance to buy a farm next to mine and I needed more power to handle that much area. I had to borrow money through the John Deere dealer to pay for the new 4020. From owning the 720, I knew the next tractor I bought had to be a diesel. That type of engine lugs so well, is economical on fuel, and lasts longer than gasoline engines.

"My 4020 had all the features of my neighbor's 4010 and then some. It had a differential lock and now delivered up to 94 horsepower. That was almost double the power of my 720 and it seemed like I could do twice as much work. I considered the optional new 8-speed power-shift transmission but decided I shouldn't borrow more money just for this convenience. After all, the standard 8-speed synchronized transmission was a big improvement over my other John Deere tractors. I did add the Roll-Gard frame after it came out in 1966. John Deere was the first tractor manufacturer to offer this roll bar to protect the operator from tractor upsets. John Deere has always been

1020 Tractor, 1965-1973, 38 hp, 3 cylinders, 8-speed transmission. Shown with a 37 Loader and 78 Rear Blade. The 1020 and 2020 utility tractors from the Dubuque factory offered many new features similar to the larger Waterloo tractors.

interested in the safety of its customers. Grandma and I were always interested in safety too, Jim.

"More than 165,000 4020 Tractors were sold in the United States and Canada from 1964 to 1972. That makes it the most popular tractor John Deere made after the 2-cylinder tractors. They still bring more at farm sales than the $10,000 I paid when it was new.

"The new tractors were especially good for loading manure. The 2-cylinder tractors were good with loaders, but the new ones were so much easier to shift gears and to maneuver. I can remember how hard my dad worked hauling manure with a horse-drawn spreader. Fortunately, this took place in the spring before I got out of school and had time to help him. The manure had to be dug out a fork full at a time to load the spreader. Several of our neighbors didn't even have a spreader, so they had to then hand-fork it off the wagon."

Neil Armstrong was the first man to step on the moon. He and Buzz Aldrin took their first walk on the moon on July 21, 1969. At that time, farming seemed completely remote from space travel.

The Sound-Gard body offered a new level of comfort and convenience for the driver of 30 series tractors. It offered a panoramic view and shielded him or her from dust, heat, and cold. Its reduced noise level made it practical to have a good radio for news, weather, and music.

1973 – All the Comforts of Home

"Grandpa, what was the next big change after your 4020 Tractor?"

"John Deere called the new Waterloo tractors brought out late in 1972 'Generation II.' Most farmers called them the 30 series. Whatever you called them, they had many important improvements. Neighbor Fred ordered a 4630 as soon as he saw that new cab, called a Sound-Gard body. When he let me drive it on a hot, dusty day with the radio playing my favorite music, I wished it was time for me to get a new tractor. He got the optional 8-speed power-shift transmission. The 4630 had a new turbocharged and intercooled engine. Turbocharging uses the exhaust gases to turn a turbine to pump more air into the engine. This results in more power, better fuel economy, and a quieter engine. At 150 horsepower, this was the most powerful tractor I had ever driven.

"A new 16-speed Quad-Range transmission was an option on the 4030, 4230, and 4430. It and the standard 8-speed transmission featured a new hydraulically-controlled wet clutch.

"It was in the '70s that John Deere dropped two options that had been very important in the past. A few 4030s and 4230s were sold with gasoline engines,

8630 Tractor, 1975-1978, 225 hp, 6 cylinders, 16-speed Quad-Range transmission. Shown with a wide drawn chisel plow in western wheat country. This is the most popular 4-wheel-drive tractor John Deere has made, with sales of more than 2,000 in 1976 and 1978 in the United States and Canada.

but by the mid-'70s, John Deere stopped making gasoline engines for farm tractors. Farmers chose diesel engines for their greater durability and better fuel economy.

"But the tricycle front end, so visible on past tractors, was disappearing. It was an option on the 4030, 4230, and 4430 but the 4630 was too heavy to support with two tires under the front end. And the tricycle option was no longer offered after the 30 series tractors were replaced in 1977. It's estimated John Deere built more than one million tricycle tractors in the 50 years they were offered, Jim.

"The '70s was a period of high farm income because we were selling lots of grain to other countries. Farmers had enough money to replace their older tractors. My brother, and many other wheat farmers in the Dakotas, bought 8630 4-wheel-drive tractors. It gave him the power he needed to pull the wide equipment used on large wheat farms. Some large farmers in the Cornbelt also bought them, mainly for tillage."

Texas Instruments sold the first commercial, electronic hand-held calculators in 1972. The TI-2500 Datamath was introduced at less than $120. Calculators helped solve the many math problems of running a farm.

2840 Tractor, 1977-1979, 80 hp, 6 cylinders, 12-speed Hi-Lo transmission. Hi-Lo permitted shifting between ranges without clutching. Shown with an 8-row front-mounted cultivator. The 2840 soon became John Deere's most popular utility tractor because it offered ample power in a small package at a reasonable price.

1978 – More Muscular Row-Crop Tractors

"Grandpa, what was the next tractor you bought after the 4020?"

"I bought a new 4440 in 1979 with a 16-speed Quad-Range transmission. This allowed me to shift down without clutching when I hit tough spots or wanted to slow down to turn at the end of the row. I especially liked the new seat suspension that smoothed my ride when disking diagonally across rough corn rows. The 40 series tractors were built more rugged, had more power, and could lift and pull more than the 30 series. The 4440 at 130 hp was

4840 Tractor, 1978-1982, 180 hp, 6 cylinders, 8-speed power-shift transmission. Power shift permitted shifting between all gears without clutching. Shown with a 770 Double Disk Air Drill. The 4840 was an added model offering more power. Engine displacement was increased on the four largest models.

the best-selling size and just right for my farming.

"The size of John Deere's best selling tractor had gone from the 35 hp 60 Tractor in 1953 to the 125 hp 4430 Tractor in 1973. This rapid increase in power was the result of increasing farm sizes and deeper tillage. In 1996, John Deere's best selling tractor is the 7800 at 145 hp. Farmers started switching to conservation tillage in the 1970s so have not needed to up their power so much for heavy tillage. Most farmers use a moldboard plow only once every few years. Many farmers have switched to no-till for most of their crops."

IBM introduced its Personal Computer in 1981. That basic design is now used in most computers throughout the world. Farmers bought computers to provide the information needed to make better farm management decisions.

"How do farmers control their weeds if they don't do a lot of tillage?"

"Farm chemicals have done wonders in protecting crops from weeds. Weeds were always a problem when I was growing up. Farm chemicals weren't available then. Dad would plow some in the fall and some in the spring. He would disk twice before harrowing to kill all the weeds. With prompt planting, he hoped the corn and beans would get a head start on the weeds and grass. We cultivated the corn at least three times before it got too tall to cultivate. We also hand hoed some of the worst patches of weeds in the corn. Many farmers check rowed their corn so they could cultivate with the rows and across the rows. Today, some farmers may not use their cultivator during years when chemicals adequately control the weeds."

"Do farmers around the world drive the same John Deere tractors we use in the United States?"

"Yes and no, Jim. Most farms in Europe are smaller, so John Deere builds many of its utility tractors, like the 2840, in Germany. But some of these are shipped to the United States and many European farmers needing larger John Deere tractors get them shipped from Waterloo."

2150 Tractor, 1983-1986, 46 hp, 3 cylinders, 8-speed synchronized transmission. Shown with 330 Round Baler. The new transmission permitted shifting on-the-go within gear ranges. This was an ideal low-cost second tractor for haying for the livestock farmer. He usually had a larger tractor for tillage and planting.

1983 – Mechanical Front-Wheel Drive

"Grandpa, how did you feed cattle before round balers were made?"

"There were many ways but none of them very easy. When I was your age, my dad owned a corn binder. Each fall the binder would bundle stalks of corn in the field. Then two men had to shock the corn. You still see pictures of pumpkins and corn shocks at Halloween. When snow was on the ground in the winter, Dad would hitch a team of horses to a sled and gather a load of bundles from the corn field. He would take it

2550 Tractor, 1983-1986, 65 hp, 4 cylinders, 8-speed synchronized transmission. Shown with 245 Loader. Mechanical front-wheel drive is especially valuable for loading manure. The ground is often slippery and the loader adds weight to the front drive wheels for improved traction.

to the cows in the pasture and unload bundles on the snow as the team walked forward. When the shocked corn ran out, he would do the same thing with hay from stacks made in the summer. Spreading the feed on the snow kept it clean and spread it out so each cow got her share.

"It seemed like John Deere had the ultimate solution to haying when it came out with the bale thrower in 1957, making one-man haying practical. But this only solved making hay rather than feeding it. The round balers John Deere started selling in1975 improved both making and feeding hay. The dairyman could use a John Deere loader to move the bales from the alfalfa field and stack them, three high, inside a shed. The beef farmer most often stores round bales in the field or near the farmstead. Both types of farmer generally use a loader to move round bales from storage into round bale feeders to minimize waste."

"Did your neighbor, Fred, buy his 4850 Tractor to move big round bales?"

"No, not really, Jim. I knew he would buy one of the 50 series tractors as soon as I saw it offered mechanical front-wheel drive. He farms so much land, he starts spring field work before the soil is really dry enough. He said he bought the 4850 because it would pull heavier loads with its greater power and mechanical front-wheel drive. It was the first MFWD he had seen that still permitted short turns at the ends of the row. He could also get more productivity from its 15-speed power-shift transmission because the speeds offered the right choice for each load. He also liked the new electronic warning system for 10 critical tractor functions. If oil pressure was too low or the engine too hot, it warned the operator by sight and by sound."

4650 Tractor, 1983-1988, 165 hp, 6 cylinders, 16-speed Quad-Range transmission. Shown with 856 Bulldozer. Mechanical front-wheel drive is needed on high horsepower tractors to get maximum drawbar pull for jobs like tillage and bulldozing.

4955 Tractor, 1989-1991, 202 hp, 6 cylinders, 15-speed power-shift transmission. Shown with 7-bottom 2810 Semi-Integral Plow. This tractor set new records in Nebraska Tests for horsepower and fuel efficiency. Most customers chose mechanical front-wheel drive to best use this power in the field.

1992 – Tough New Utility Tractors

"Grandpa, has John Deere made any completely new farm tractors since they switched from the 2-cylinder tractors in the '60s?"

"Yes, they started replacing the entire farm tractor line in 1992 but they kept the same basic engines. They even built a new factory in Augusta, Georgia, to make 40- to 60-hp tractors. These tractors are well matched to the Southeast where farms are smaller than in the Cornbelt or farther west. These 3-cylinder tractors have been quite popular so they added a fourth model, the 4-cylinder 5500 with 73 hp. Engines for these tractors come from another John Deere factory in France. You can also get a 12-speed synchronized transmission with power reverser that requires no clutching to change directions. The air-conditioned cab is also a good option."

5200 Tractor, 1992- , 41 hp, 3 cylinders, 9-speed transmission. Shown with 603 Rotary Cutter. This was the first complete redesign of utility tractors since the 1020 and 2020 were introduced in 1965. These rugged new tractors had many standard features that made them a great buy in this size class.

5400 Tractor, 1992- , 60 hp, 3 cylinders, 9-speed transmission. Shown with 540 Loader. Livestock farmers like the 5400 as their second tractor for its maneuverability. The tractor and loader are delivering a bale to a round bale feeder. The concrete feed bunk behind the tractor also serves as a fence.

"Grandpa, it seems like everything on the farm can now be done with the help of a tractor or a combine. What was the last part of farming to be mechanized?"

"The feeding of livestock was the last work on farms to be mechanized. We used to have chickens, sheep, hogs, horses, and lots of cattle on our farm. Most farms had livestock when I was your age. I used to turn a corn sheller by hand to shell corn for the chickens. We shoveled ear corn out on the ground for our hogs to eat. We carried oats in a bucket to feed each horse in his own trough. We either chopped or broke the ears of corn we fed our cows and calves. We threw hay down from the barn loft with a pitch fork to feed our horses and cattle. Then we had to carry it to their mangers.

"John Deere sold several machines that made processing of feed a bit easier. Power shellers and hammer mills were popular in the '50s and '60s. Grinder-mixers were popular during the late '60s and the '70s to grind feed grains where they were stored, to mix them with supplements, and to deliver them to hog or cattle pens. The last big change for feeding grain and silage to cattle came with the use of fence-line concrete feed bunks. These kept the cattle on one side and the feed could be delivered continuously by driving along the other side. Forage wagons were popular for this in the '60s, '70s, and '80s but many have been replaced by mixer-feeder wagons."

6200 Tractor, 1992-, 66 hp, 4 cylinders, 12-speed synchronized transmission. Shown with an 820 Sicklebar Mower Conditioner. The all-new 6000 series tractors are well matched to the many operations on beef and dairy farms. They are large enough for tillage and planting, but economical enough for haying.

7800 Tractor, 1992-, 145 hp, 6 cylinders, 16-speed PowrQuad transmission. Gear selections within ranges require no clutching. Shown with a 750 No-Till Drill. The all-new 7800 with MFWD is currently John Deere's most popular tractor. It is the size needed by many corn and soybean farmers.

1992 – An All-New Breed of Power

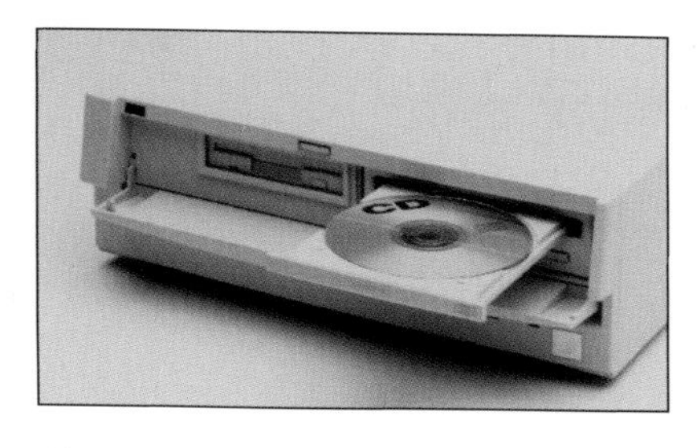

IBM started offering CD-ROMs as an option on some of their personal computers in 1993. Farmers are trading up their computers to provide the information needed to make timely marketing decisions.

I'm doing a Soil and Water Conservation project in my 4-H club. My mom is the 4-H leader for this and several other projects the members are working on in our club. I get a lot of help on the project from my dad because he knows even more than Grandpa Turner about soil conservation. Dad has been on the Soil and Water Conservation District Board for several years. "Dad, what are some of the changes in soil conservation since you started farming in 1975?"

"The county extension service was already promoting reduced tillage in the '70s, but the university researchers still had a lot to learn about the best way to conserve soil and still get high yields in our area. It has taken the work of many groups to make the various practices as effective as they are today, Jim. We knew that leaving last year's crop residue or trash on top of the soil reduced erosion and saved water. However, this encourages more weeds and insects. The chemical companies have done a good job in providing us with selective chemicals to protect our plants.

"John Deere has developed a variety of tillage implements to leave much of the trash on top. These include V-Rippers, Disk Rippers, Mulch Tillers, Mulch Masters, Mulch Finishers, Chisel Plows, and Field Cultivators. They also developed other implements needed for reduced tillage with more trash left on the surface. The most popular one in our area is the 750 No-Till Drill for soybeans and wheat. The 7200 MaxEmerge 2 Conservation Planter is equally important for corn. John Deere offers several types of row-crop cultivators. The Hi-Cycle Sprayer is often chosen by custom operators."

Dad and Mom are also active members of Farm Bureau. Mom read in one of its folders that one farmer in the United States now grows enough food to keep 129 people fed. That is about as many as we have in our grade school or go to our church. She said that 28 of these people live in other countries and eat the wheat and other foods we export. Our farmers couldn't do this if they didn't have efficient equipment and large farms.

8400 Tractor, 1995-, 225 hp, 6 cylinders, 16-speed power-shift transmission. Shown with 510 Disk Ripper. John Deere again raised the top power available in a row-crop tractor. The 510 Disk Ripper is one of John Deere's most popular conservation tillage tools.

1995 – 225 Horsepower in a Row-Crop Tractor

My mom says our new 8300 Tractor (shown on the back cover with Mom in the cab) is the easiest tractor to operate of any she's ever driven. She should know because she learned to drive on Grandpa Turner's 4020 and has driven all his other tractors, including his restored Model "D" Tractor. The 8300's right armrest swivels and floats with the seat so controls are always at the operator's fingertips. All the most used controls are there: the throttle, transmission, PTO, 3-point hitch, and three hydraulic

The patented design of the 8000 series tractors solved the two major problems with large row-crop tractors. Locating instruments and controls away from the dash permitted a "wasp waist" for excellent vision. The engine is high enough for the front wheels to fit under it for tight turns when cultivating 30-inch wide rows.

functions. The control panel to the right of the seat even has a new engine control to hold speed like cruise control does on cars.

"Dad, why did you buy the new 8300 Tractor last year?"

"We rented another farm and needed more power for tillage, planting, and cultivation. I like row-crop tractors better than 4-wheel-drive tractors because they are more maneuverable and easier to get on and off. Its ideal for cultivation because it has good vision and short turning at the ends of the field. And with the new Row-Trak guidance system I got, it's a lot easier to keep the rear-mounted cultivator on the row."

I always wondered how Grandpa Turner knew so much about John Deere equipment so I asked him one day, "Grandpa, how come you know so much about John Deere?"

"Well, I read a lot, Jim. Here are some of the books I have. *The Toy and the Real McCoy* is probably the best one for you to start on. I like *How Johnny Popper Replaced the Horse* because it tells even more about the 2-cylinder tractors that I grew up with. When I want accurate information on serial numbers and horsepower, I refer to *John Deere Tractors 1918-1994*. I learned the most from the two large volumes, *John Deere Tractors and Equipment*. Your younger sister Karen already enjoys *Johnny Tractor and His Pals*, *Corny Cornpicker Finds a Home*, and *A Tractor Goes Farming*. I also read a couple of good magazines on John Deere, *Two-Cylinder* and *Green Machine*. I hope you like reading as much as I do, Jim."

I'm going to read a lot and try to learn as much about John Deere tractors and farming as Grandpa Turner and Dad know.

Tractor design has added many features for productivity and convenience since the Model "D" shown on page 10. The driver has windows on all sides for excellent visibility. The diesel engine is electronically controlled to provide extra power when needed for tough going.

The Furrow magazine is more than 100 years old. It is mailed to 1.6 million farmers, making it the most widely read farm magazine in the world. It is published in 11 languages for farmers in more than 40 countries. It encourages improved farming methods and provides information on John Deere equipment.

Seats are filled for the John Deere show several times each day at the annual Farm Progress Show. The most widely attended farm show in the world rotates between Iowa, Illinois, and Indiana. Farm families attend to learn about the latest in farm equipment, hybrid seeds, farm chemicals, and farm practices.

1996 – Learning More About John Deere

I look at and read some of my parent's farm magazines. The one I like best is *The Furrow*. It has the latest John Deere equipment in it. I especially like to read about farming that is different from ours in the Midwest. Sometimes it even tells about farming in other countries. Some of these articles help me understand some of Grandpa's stories about farming in the olden days. *The Furrow* even has some jokes and cartoons. *Implement & Tractor* is also an interesting magazine because it shows farm equipment from other companies as well as from John Deere.

The best show I ever attended was the Farm Progress Show. John Deere had a huge display of equipment and an exciting show about new equipment. We rode out to a field where Dad and Mom drove new John Deere tractors. We also saw demonstrations of forage harvesters, corn combines, soybean combines, and tillage tools used after harvest. Farmers could watch several brands working side-by-side in the same fields.

People in dealer parts departments are well trained and equipped to meet their customer's needs for repair parts information. Special publications for parts that normally wear out from use are available. The computer with CD-ROM has replaced paper parts catalogs for finding part numbers.

I even learn more about John Deere when Mom takes Karen and me with her when she goes to the John Deere dealer to get parts. I don't know how the parts men kept track of all their part numbers before they had computers. Sometimes the parts man doesn't even look up the number if Mom describes the part she needs. He just brings the part out and Mom can tell if it is the right one. The dealer keeps a record of all the John Deere equipment my parents own.

I asked the parts man to show me my sister's riding tractor on the computer screen. He showed it to me and then ran a copy on his printer, showing all the parts. It will help me whenever I need to repair her tractor. I already adjusted the seat farther back for Karen because her legs are getting longer.

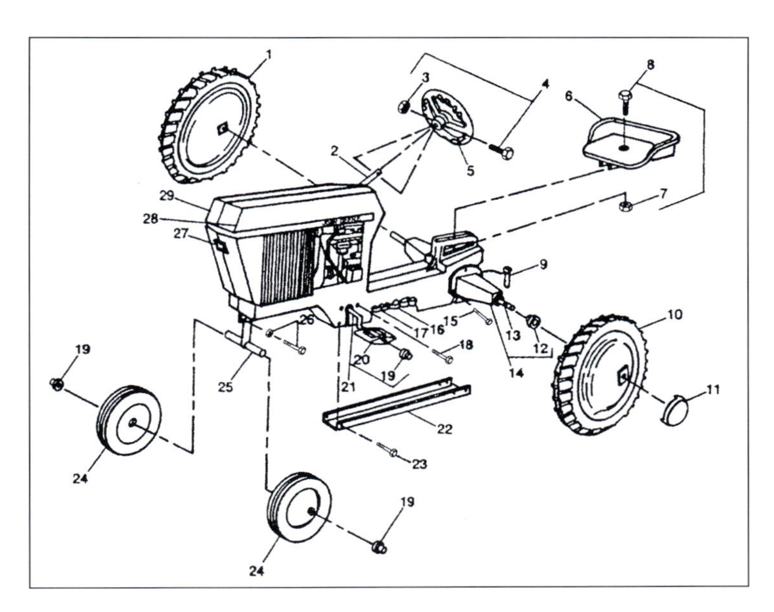

Exploded views of parts, such as for this children's riding tractor, can be shown at the top of the computer screen and the part numbers lists at the bottom. Thus, it is easy for the customer and the person in parts to see and agree on the part that is needed. This information may also be printed out as shown here.

7600 Riding Tractor made by Ertl in Dyersville, Iowa. 1 kid power, not furnished.

LX188 Lawn Tractor made in Horicon, Wisconsin. 17 engine hp, Kawasaki gasoline engine made in Japan. Shown with mower made in Horicon.

955 Compact Utility Tractor made in Horicon, Wisconsin. 27 PTO hp, Yanmar diesel engine made in Japan. Shown with snow blower made in Horicon.

5300 Utility Tractor made in Augusta, Georgia. 50 PTO hp, John Deere diesel engine made in Saran, France. Shown with 520 Loader made in Welland, Ontario.

I got to go to John Deere Day at our dealership this year with Dad and Mom. Grandpa Turner said he and Grandma have been going to John Deere Day for about 50 years. All of our neighbors were there. The dealership was all cleaned up and decorated like a birthday party. They had lots of tractors and implements on display.

6400 Tractor made in Mannheim, Germany.
85 PTO hp, John Deere diesel engine made in Saran, France.

7800 Tractor made in Waterloo, Iowa.
145 PTO hp, John Deere diesel engine made in Waterloo. Shown with 9500 Combine made in East Moline, Illinois.

8300 Tractor made in Waterloo, Iowa.
200 PTO hp, John Deere diesel engine made in Waterloo. Shown with 12-row rear-mounted cultivator made in Ankeny, Iowa.

8970 4-Wheel-Drive Tractor made in Waterloo, Iowa. 339 PTO hp, Cummins diesel engine made in Columbus, Indiana.

We saw lots more new equipment in the movies. I thought the tractors we saw there (shown on these two pages) were all that John Deere made. Dad says they are less than one-third of the tractor models and sizes that John Deere sells. We had free pop and doughnuts when we went in and they fed us lunch when the movies were over.

A 7800 Tractor nears completion on the assembly line at Waterloo. The engine, hydraulic controls, and other systems are checked at this point.

1996 – What Does the Future Hold?

This summer, Grandpa took me to the Expo at Grundy Center, Iowa. We saw hundreds of old 2-cylinder John Deere tractors and some implements. Most of them were restored and fresh painted to look brand new, like Grandpa's "D" Tractor. Mom and Dad also took me to the Midwest Old Threshers show at Mt. Pleasant, Iowa, around Labor Day. We saw several old steam engines running that Grandpa had talked about. I wouldn't want to drive one of them. They look too hot.

I wish I could have been with my Uncle Kent when he got to go through the assembly plant at Waterloo and see his new 7800 Tractor assembled on the

Precision Farming may have as much impact on the future of farming as hybrid seeds and commercial fertilizers had on the past. A global positioning satellite is shown providing the exact location within the field for a planter, a sprayer, and a combine.

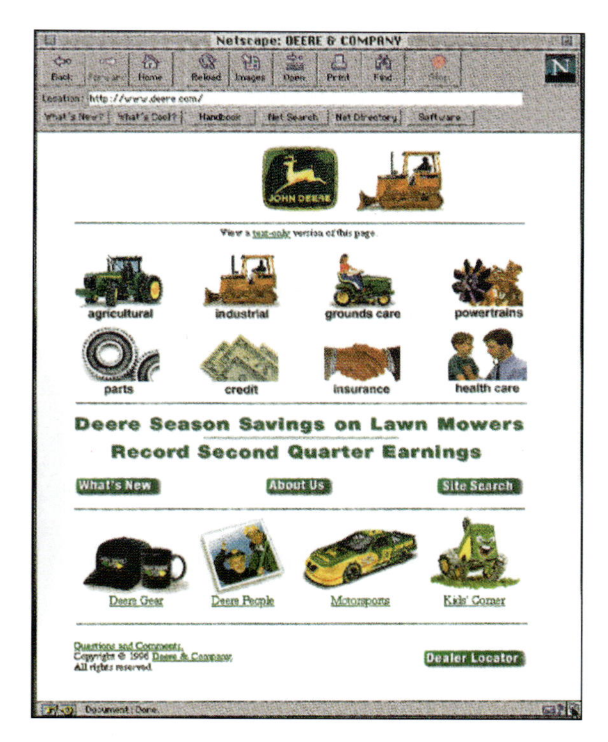

Deere & Company joined the World Wide Web in May 1996 with the above home page. Product, business, and other information are available. Deere's Internet address is **http://www.deere.com**.

line. Dad says I will be old enough to go through the factory when we get our next tractor so he and I can see it put together and tested. I can hardly wait.

"Dad, what new features do you think our next tractor will have?"

"I don't know, Jim, but I do know John Deere is a leader in new technology to help farmers. Precision Farming looks like the next big boost to farming. I have ordered the GreenStar Yield Mapping System for our 9500 Combine. As I drive through the field, it will show me yields and moisture readings on-the-go. At the end of the day I pull out a card that is read by our computer. Then I can print out a map on our color printer, showing yield levels throughout the field. Some of the commercial fertilizer applicators can already vary fertilizer rates across the field, based on soil tests and yield maps. Planters might vary seeding and fertilizer rates for higher yields. Sprayers could target chemical application to weedy areas only, to help the environment and reduce costs."

I still can't imagine what tractors in the future will be like. Look at the big jumps in design between the steam engines, Grandpa's Model "D", his 4020, and now Uncle Kent's 7800 and our 8300. I am going to keep up-to-date on John Deere as soon as Mom or Dad shows me how to reach it on the Internet. The John Deere site shows lots of equipment photos and specifications.

I don't know if I want to be a farmer and use John Deere's future tractors or maybe I should be an engineer. Then I could help design their future tractors — I wonder if I can get chocolate-chip cookies at John Deere.

Glossary

1917-1924 — Tractor production years. Like new car models, tractors may be sold in the latter part of the preceding year.

Deere & Company — The name of the parent company of John Deere factories, engineering centers and sales branches.

Depression — A period of widespread unemployment and low income.

Drawbar — The bar on the tractor used to attach and pull drawn implements.

Horsepower — A measure of the rate of work a tractor can do. Values given for 2-cylinder tractors are rated horsepower measured at the belt pulley. Tractor power from 1959 on is maximum horsepower measured at the power take-off (PTO). One horsepower (hp) can lift 550 pounds one foot in one second. A 100 pound person climbing up stairs 10 feet in 3.6 seconds produces .5 horsepower.

http://www.deere.com — Home page address for Deere & Company.

Hybrid Seed Corn — Hybrid seed results from crossing two unlike varieties to give higher yields. Farmers have to purchase hybrid seed each year instead of saving their own seed as in the past.

REA — The Rural Electrification Administration is a cooperative started by the government to provide electricity to farmers.

Sales Branches — The organizations that provide sales and service to John Deere dealers.

Sulky Rake — A rake that gathers and then dumps hay into windrows.

Transmission — The set of gears used to transmit power from the engine to the drive wheels. All 2-cylinder tractors changed speeds and direction by sliding gears on splined shafts and thus had to stop to shift gears. Many utility tractors since then have constant-mesh gears but still need to stop to change gears. Most larger tractors used synchronized gears to permit shifting on-the-go by clutching. Power shifts can be made between some gears or ranges on Quad-Range, Hi-Lo, and PowrQuad transmissions. Full power-shift transmissions can be shifted through all gears on-the-go without clutching.

Index

About the Author

Roy Harrington grew up on a livestock and grain farm. He worked more than 30 years as an engineer planning and developing John Deere farm equipment. Roy is co-author of *John Deere Tractors and Equipment 1960-1990*, a best seller farm equipment history book. He is also author of a children's book, *A Tractor Goes Farming*. His five grandchildren are fascinated by the three tractors he drives. His oldest tractor is a 1947 Model "A" John Deere.

About ASAE — The American Society of Agricultural Engineers

ASAE is a technical and professional organization committed to improving agriculture through engineering. Many of our 8,000 members in the United States, Canada, and more than 100 other countries are engineering professionals actively involved in designing the farm equipment that continues to help the world's farmers feed the growing population. We're proud of the triumphs of the agricultural and equipment industry.